RAND | Homeland Security and Defense Center

T0097337

Capabilities-Based Planning for Energy Security at Department of Defense Installations

Constantine Samaras, Henry H. Willis

This monograph results from the RAND Corporation's continuing program of self-initiated research. Support for such research is provided, in part, by donors and by the independent research and development provisions of RAND's contracts for the operation of its U.S. Department of Defense federally funded research and development centers.

Library of Congress Cataloging-in-Publication Data

Samaras, Constantine.
  Capabilities-based planning for energy security at Department of Defense installations / Constantine Samaras, Henry H. Willis.
    pages cm
  Includes bibliographical references.
  ISBN 978-0-8330-7910-7 (pbk. : alk. paper)
  1. United States. Dept. of Defense—Energy consumption. 2. United States. Dept. of Defense—Energy conservation. 3. Strategic planning--United States. 4. Energy policy—United States. I. Willis, Henry H. II. Title.

HD9502.U52S25 2013
355.7068'2—dc23

                                                              2013003796

Cover photo courtesy of NASA: NASA Earth Observatory/NOAA NGDC.

Published 2013 by the RAND Corporation
1776 Main Street, P.O. Box 2138, Santa Monica, CA 90407-2138
1200 South Hayes Street, Arlington, VA 22202-5050
4570 Fifth Avenue, Suite 600, Pittsburgh, PA 15213-2665
RAND URL: http://www.rand.org/
To order RAND documents or to obtain additional information, contact
Distribution Services: Telephone: (310) 451-7002;
Fax: (310) 451-6915; Email: order@rand.org

Preface

Since Department of Defense (DoD) installations in the United States rely on the commercial electricity grid for 99 percent of their electricity needs, nearly all critical functions on installations depend on infrastructure outside DoD's control. In a large-scale complex disaster or terrorist attack, the installations would be a base of operations for emergency services. Thus, the 2010 Quadrennial Defense Review identified diversifying energy sources and increasing efficiency in DoD operations as critical goals. Using portfolio analysis methods for assessing capability options, we discuss an energy security framework to assist DoD in evaluating choices among portfolios of energy technologies to maintain adequate power to its critical missions located in the United States, tradeoffs among capabilities, and cost among the portfolios.

This monograph results from the RAND Corporation's continuing program of self-initiated research. Support for such research is provided, in part, by donors and by the independent research and development provisions of RAND's contracts for the operation of its U.S. Department of Defense federally funded research and development centers.

The RAND Homeland Security and Defense Center

The research reported here was conducted in the RAND Homeland Security and Defense Center, a joint center of the RAND National Security Research Division and RAND Justice, Infrastructure, and Environment. RAND Justice, Infrastructure, and Environment provides insights and solutions to public- and private-sector decision-makers across numerous domains, including criminal and civil justice; public safety; environmental and natural resources policy; energy, transportation, communications, and other infrastructure; and homeland security. RAND Justice, Infrastructure, and Environment studies are coordinated through four programs—the Institute for Civil Justice; the Safety and Justice Program; the Environment, Energy, and Economic Development Program; and the Transportation, Space, and Technology Program—and the Homeland Security and Defense Center, run jointly with the RAND National Security Research Division. The RAND National Security Research Division conducts research and analysis for all national security sponsors other than the U.S. Air

Force and the Army. The division includes the National Defense Research Institute, a federally funded research and development center whose sponsors include the Office of the Secretary of Defense, the Joint Staff, the Unified Combatant Commands, the defense agencies, and the U.S. Department of the Navy. The National Security Research Division also conducts research for the U.S. intelligence community and the ministries of defense of U.S. allies and partners. The Homeland Security and Defense Center conducts analysis to prepare and protect communities and critical infrastructure from natural disasters and terrorism. Center projects examine a wide range of risk-management problems, including coastal and border security, emergency preparedness and response, defense support to civil authorities, transportation security, domestic intelligence, and technology acquisition. Center clients include the U.S. Department of Homeland Security, the U.S. Department of Defense, the U.S. Department of Justice, and other organizations charged with security and disaster preparedness, response, and recovery.

Questions or comments about this report should be sent to the project leader, Constantine Samaras (Constantine_Samaras@rand.org). For more information about the Homeland Security and Defense Center, see http://www.rand.org/hsdc or contact the director at hsdc@rand.org.

Contents

Figures

Tables

Summary

Extensive energy delivery outages in 2012, such as the widespread electricity, natural gas, and refined oil product disruptions due to Hurricane Sandy; the summer weather-related outages in the Washington, D.C., area; and the largest blackout in global history in India, have reinforced public and policymaker awareness of risks to the electricity infrastructure system. The U.S. electricity grid is vulnerable to disruptions from natural hazards and actor-induced outages, such as physical or cyber attacks.

Since Department of Defense (DoD) installations in the United States rely on the commercial electricity grid for 99 percent of their electricity needs, nearly all critical functions on installations depend on infrastructure outside DoD's control. In the event of a catastrophic disaster—such as a severe hurricane, massive earthquake, or large-scale terrorist attack—DoD installations would be a base for emergency services. The 2010 Quadrennial Defense Review therefore identified diversifying energy sources and increasing efficiency in DoD operations as critical goals. In response, DoD is developing installation energy security portfolios to enhance energy security that use a mix of new technologies and modifications to operations.

Currently, the notion of enhancing energy security on DoD installations is not fully defined. Energy security for how long? Under what conditions? At what cost? And most importantly, for what reasons? Without an understanding of these issues, planning for, executing, and evaluating proposed enhancements is challenging. The underlying analytical question for energy security is, "What critical capabilities do U.S. installations provide, and how can DoD maintain these capabilities during an energy services disruption in the most cost-effective manner?" Answering this question requires a systems approach that incorporates technological, economic, and operational uncertainties. In other words, this problem is well suited for capabilities-based planning (CBP)—planning under uncertainty that (if done well) provides capabilities for a wide range of challenges within economic constraints.

Capabilities-based planning means different things to different people, and some aspects of its implementation in DoD have been appropriately controversial. In this report, we have in mind the core feature—planning under uncertainty—to provide capabilities for a wide range of challenges (including diverse circumstances) while working within economic constraints. This means making choices in allocating limited resources to be best able to deal with future demands, which cannot be perfectly anticipated or defined in advance. In this context, *capabilities* includes the broad meaning associated with "ability and wherewithal," not just "assured ability to do a very specific task in a very specific set of circumstances." Further, this form of planning is inherently about making choices; it is the opposite of a blank-check approach. Finally, despite confusion on the matter a decade ago, capabilities-based planning also includes using concrete scenarios to test the effectiveness of options. However, such scenarios should be chosen analytically to be good test cases of broad capability, with no illusions about their being meaningful "best estimates." This interpretation of capabilities-based planning is quite consistent with the most recent Quadrennial Defense Review.

In this short report, we use RAND's portfolio analysis methods for assessing capability options to present the foundation of a framework that could be developed to evaluate energy security portfolios for these installations. While capabilities-based planning is now a component of DoD decisionmaking, we discuss how it might be extended and used for energy security planning. We outline an approach with some examples and detail, but these should be viewed as scoping suggestions—a fully developed DoD planning framework would incorporate mission context and relevant current issues. Our intention with this think piece is to stimulate a discussion of how DoD installation energy services contribute to homeland defense and homeland security, how DoD can evaluate choices to maintain adequate energy services to critical missions located in the United States, and how DoD can make tradeoffs among capabilities and costs to maintain these capabilities during disruptions.

DoD Could Use Joint Capability Areas (JCAs) as the Foundation for Measuring Energy Security

DoD analysts planning for installation energy security rely on the broad energy security definition established in the 2012 National Defense Authorization Act—assured energy access, reliable supplies, and sufficient energy to meet mission essential requirements. The remaining challenges for decisionmakers include measuring these terms, defining and ranking mission essential requirements, and making tradeoffs among costs and capabilities. Each of these can and should be undertaken for each installation, but a broader, DoD-wide capabilities assessment could inform decisionmakers on maintaining overall capabilities during the loss of energy services.

DoD has divided the department's capabilities into functional categories (JCAs) to enable capabilities-based planning for warfighting needs. These Joint Capability Areas are Force Support, Battlespace Awareness, Force Application, Logistics, Command and Control, Net-Centric, Protection, Building Partnerships, and Corporate Management and Support. Although the current Joint Capability Areas are clearly written to codify many capabilities required for homeland defense, their applicability in supporting homeland security functions in response to a natural disaster or terrorist attack is less clear. Portions of the existing JCAs do have language about providing essential services and protecting assets throughout disasters, but DoD would potentially need to refine the JCAs to further emphasize capabilities required for civil support. Yet, existing Joint Capability Areas can be used to demonstrate capabilities-based metrics for installation energy security analyses. The nature and extent of additional DoD capabilities required for civil support that are enabled by installation energy could be defined during a fully developed analysis.

To demonstrate an assessment of DoD capabilities, we use Joint Capability Areas to form the foundations of metrics to evaluate installation energy security decisions. In essence, we ask if existing or proposed installation energy security strategies enhance DoD capabilities, and we evaluate strategy cost-effectiveness. To define the capabilities used for an energy security analysis to be the most relevant for decisionmakers, we propose top-level, plain English functions across each Joint Capability Area. We define these functions to be the provision of training, information, materiel, care, and security. To assess effectiveness across each function under various energy security strategies, several analytic measures of effectiveness would underlie the evaluation of each function.

DoD Could Construct a Broad Scenario Space to Evaluate Energy Security Strategies

In planning, it is essential to consider a broad range of future challenges—what is often referred to as a broad "scenario space," with the word *scenario* referring not just to a generic category, such as a natural disaster, but to a specific example with all of the details that fully define it. An infinite number of such scenarios exist, so—after thinking about the broad scenario space—planners need a smaller set of test cases with which to work. These tests can only be illustrative but should be chosen analytically to stress the options under consideration in all the important dimensions. The resulting set of test cases has been called a "spanning set" to suggest that an option that does well in all of the test cases should be able to do well in a real world case, even though that would most likely be different from any of the test cases.

A starting point for developing a spanning set of test cases for evaluating energy security is defining the phenomena that lead to energy disruptions. Installations predominately derive energy services through use of energy commodities produced by others externally and delivered to the installation via public infrastructure. These include the commercial electricity fuels extraction, transportation, production, transmission, and distribution systems; the commercial natural gas extraction, processing, storage, transmission, and distribution systems; and the global and national markets for crude oil extraction, transportation, refining, storage, and distribution systems.

Each node across these supply chains represents an opportunity for accidental or intentional disruption, which could potentially affect an installation's ability to receive these commodities as expected. Additionally, installation-distributed generating assets themselves, as well as their enabling infrastructure, remain vulnerable to disruption. This characterization defines a spanning set of test cases, illustrated in Figure S.1, that includes four classes of disruptions: (1) a loss or reduction of electricity from the commercial grid, (2) a loss or reduction of natural gas from the commercial distribution system, (3) a disruption of petroleum resupply to an installation, and (4) the loss or availability reduction of energy assets within an installation. The four classes should be analytically expanded into test cases along the dimensions of complexity, scale, disruption time, preparedness, and response resources. Using a common framework and systems approach, DoD can link and evaluate how energy security tasks and strategies affect DoD-wide Joint Capability Areas during a loss of installation energy services across the test cases constructed.

Using Capabilities-Based Planning to Inform Decisionmaking About Installation Energy Security

The examples described in this report demonstrate how capabilities-based planning could be used to inform choices about the adoption of technologies and practices to

Figure S.1
Four Spanning Test Cases Illustrating Pathways for Installation Energy Services Disruption

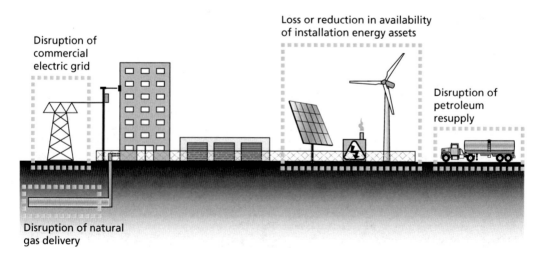

NOTE: Illustrative and not to scale.
RAND RR162-S.1

enhance energy security. In doing so, the report demonstrates the steps involved in this analysis and the types of data that are needed.

DoD can use a diverse set of technologies and strategies to enhance installation energy security. This large set might include locally generated renewable energy, implementation of emergency load management plans, purchasing energy storage assets, and many other choices. Using the language of previous RAND work on capabilities-based planning, we call each of the technologies and strategies for energy security a "building block." We characterize the sample building blocks into three groups: (1) efficiency, operations, and information investments; (2) energy generation, distribution, and control capital equipment; and (3) fuels, energy storage, and enabling equipment capital purchases. The performance of each building block would depend on assumptions about local resources and the technical and economic characteristics of each building block as well as the presence and synergies of other building blocks (e.g., microgrid availability and renewable electricity). After screening the landscape of building block combinations, several composite options could be chosen for analysis, each made up of a set of energy security building blocks. An analyst generating composite options with energy assets and strategies would necessarily need to incorporate the uncertainty of current and future costs and performance for each building block to fully represent the range of possibilities involved with each composite option. Energy security portfolios should be evaluated in terms of performance and effectiveness in maintaining the provision of training, information, materiel, care, and security over the test cases and their dimensions. This allows performance of energy security strategies to be evaluated *across* test cases, such as the loss of electricity and natural gas. Developing an analytical hierarchy of measures of effectiveness across test cases and dimensions allows decisionmakers to drill down and understand the drivers for shortfalls. These shortfalls could result from technical, economic, resource availability, or other reasons. DoD decisionmakers can then use the sensitivity analysis ranges established in the estimates to understand how underlying variables, such as technology cost and performance improvements, can affect outcomes.

For many applications, the thresholds revealed by these sensitivity analyses could inform technology cost and performance targets and decisionmaking across the DoD research, development, test, and evaluation (RDT&E) enterprise. For example, an estimated cost target for energy storage on a per-kilowatt-hour of energy basis that would enable greater cost-effectiveness of maintaining DoD capabilities could emerge from a capabilities-based portfolio analysis and inform RDT&E portfolio managers. Thus, broader opportunities can be identified with a capabilities-based analysis. Finally, these analyses can help researchers identify surplus capabilities, such as multiple, oversized diesel generators not integrated into microgrids. DoD could consider reducing the identified surpluses in one capability area to save resources for capability enhancements in other areas.

Over a range of decisionmaker perspectives, the use of cost-effectiveness landscapes generated from capabilities-based portfolio analyses can assist DoD in making choices about installation energy security strategies, both on individual installations and across the DoD's portfolio. The process is analytically intensive, yet it can reveal weaknesses and vulnerabilities in existing and proposed strategies to maintain installation energy services. By defining and evaluating a data-driven portfolio approach and incorporating the associated uncertainties, the DoD can better prepare for unexpected energy disruptions at installations that affect both homeland defense and homeland security.

Acknowledgments

This work was performed within the RAND Corporation's continuing program of self-initiated research. Support for such research is provided, in part, by donors and by the independent research and development provisions of RAND's contracts for the operation of its U.S. Department of Defense federally funded research and development centers. We are grateful to the RAND Homeland Security and Defense Center for supporting this research. We thank Sandra Petitjean for graphics production. The invaluable comments received from our peer reviewers, Paul Davis and John Matsumura, greatly improved this document.

Abbreviations

BA	battlespace awareness
C4ISR	command, control, communications, computers, intelligence, surveillance, and reconnaissance
CBP	capabilities-based planning
CONOPS	concept of operations
CONUS	continental United States
DoD	Department of Defense
DOE	Department of Energy
DOTMLPF	Doctrine, Organization, Training, Materiel, Leadership and Education, Personnel, and Facilities
EPA	Environmental Protection Agency
FERC	Federal Energy Regulatory Commission
GAO	Government Accountability Office
IPCC	Intergovernmental Panel on Climate change
JCA	Joint Capability Area
JDDE	joint deployment and distribution enterprise
JFC	joint force commander
LE	law enforcement
MTS	Maneuver to Secure
NC	Net-Centric
NERC	North American Electric Reliability Corporation

NRC   National Research Council

PAT   Portfolio Analysis Tool

P.L.   Public Law

PNT   Position, Navigation, and Timing

PV   photovoltaic

RDT&E   research, development, test, and evaluation

CHAPTER ONE

Introduction

Since Department of Defense (DoD) installations in the United States rely on the commercial electricity grid for 99 percent of their electricity needs, nearly all critical functions on installations depend on infrastructure outside DoD's control (Stockton, 2011a). Beyond their normal military functions, these installations would be a base for emergency services after some natural or human-caused disasters. The 2010 Quadrennial Defense Review therefore identified diversifying energy sources and increasing efficiency in DoD operations as critical goals (DoD, 2010), and energy security has reemerged as a priority throughout the department. As one aspect of this, DoD is evaluating various strategies to enhance energy security at installations. But how can installation energy security be specifically defined and measured? How does DoD's role in supporting civilian agencies during disasters shape installation energy decisions? What should these energy security strategies look like? How can they be assessed and implemented cost-effectively? In this short report, we use RAND's portfolio analysis methods for assessing capability options to present the foundation of an installation energy security framework to potentially answer these questions.

DoD installation energy security decisions are well suited for capabilities-based planning (CBP)—planning under uncertainty that (if done well) provides capabilities for a wide range of challenges within economic constraints. For installation energy security analyses from a DoD-wide perspective, the objective should be establishing valid and appropriate measures. The core question in establishing these measures should be: Which DoD capabilities are enabled by energy services and which capabilities are required even when traditional power sources are unavailable? DoD can then evaluate potential options using practical, reliable, and transparent metrics. In other words, how much additional capability is gained by adopting a specific portfolio of technologies and strategies? Conducting good capabilities-based planning is about informing and explaining tradeoffs and choices and illustrating how an organization makes judgments about aggregating the associated risks and benefits.

Although capabilities-based planning is now a component of DoD decisionmaking, analysts could extend its use for energy security planning. We outline an approach with some examples and detail, but these should be viewed as scoping suggestions—a fully developed DoD planning framework would incorporate mission context and relevant current issues. Our intention with this think piece is to stimulate a discussion of how DoD installation energy services contribute to homeland defense and homeland security, how DoD can evaluate choices to maintain adequate energy services to critical missions located in the United States, and how DoD can make tradeoff decisions to maintain these capabilities during disruptions.

Background

DoD operates more than 4,400 domestic sites[1] spanning all 50 states and the District of Columbia. Although about 10 percent of domestic DoD sites are land holdings without any facilities, more than 200 DoD sites are medium or large operations, each with an estimated replacement value of more than about $1 billion. Nearly all of these sites require energy to function. Installations provide overall military capabilities not with the actual electricity, natural gas, or petroleum needed for operations but with the energy services provided by these commodities: lighting, heating, communications, refrigeration, food preparation, computing, and other services. A loss of energy services at an installation affects the installation's ability to perform specific mission capabilities. During the current contingency operations, tactical unmanned aircraft systems in theater are sometimes piloted from U.S.-based installations (Bumiller, 2012). Furthermore, many U.S. installations have enhanced command, control, communications, computers, intelligence, surveillance, and reconnaissance (C4ISR) capabilities supporting highly critical missions. Hence, there is overlap between what are traditionally thought of as *installation energy* and *operational energy* needs. Yet, nearly all U.S. installations depend on the commercial electricity grid and natural gas distribution system.

A 2008 Defense Science Board report identified four sources of risk for loss of power at installations: grid failure from overload, destruction from natural disasters, terrorist attacks and sabotage, and cyber attacks (Defense Science Board, 2008). These events pose different levels of risks depending on severity and location. The complex and sprawling nature of the electricity system creates vulnerabilities that enable the possibility that determined actors could disable large potions of the electricity grid for weeks or perhaps months (National Research Council [NRC], 2012). Yet installations currently rely on diesel generators with short-term fuel stockpiles for backup power during outages. Stockton (2011b) noted that DoD has become more dependent on civilian systems, such as transportation, energy, information, and commerce, even as these systems become more at risk of natural and actor-induced disruption. DoD energy security planning therefore needs to consider this civil-military symbiosis.

---

[1]   Within the domestic sites, DoD manages a total of 238,164 buildings, 164,986 structures, and 37,353 linear structures (DoD, 2013).

The 2012 National Defense Authorization Act redefined DoD energy security as "having assured access to reliable supplies of energy and the ability to protect and deliver sufficient energy to meet mission essential requirements" (Public Law [P.L.] 112-81, 2011). Furthermore, it directs the Secretary of Defense to establish a policy that provides:

(A) Favorable consideration for energy security in the design and development of energy projects on the military installation that will use renewable energy sources.

(B) Guidance for commanders of military installations inside the United States on planning measures to minimize the effects of a disruption of services by a utility that sells natural gas, water, or electric energy to those installations in the event that a disruption occurs.

Extensive energy delivery outages in 2012, such as the widespread electricity, natural gas, and refined oil product disruptions due to Hurricane Sandy; the summer weather-related outages in the Washington, D.C., area; and the largest blackout in global history in India, have reinforced the public's awareness of risks to the electricity infrastructure. The U.S. electricity grid is vulnerable to disruptions from such natural hazards as weather and earthquakes; unplanned outages from equipment failure, error, mismatches in supply and demand, or accidents; and actor-induced outages, such as physical or cyber attacks (Stockton, 2011a; Eaton Corporation, 2012). The dominant causes of disruptions are weather-related (Department of Energy [DOE], 2012a), raising the importance of planning and policy contingencies for potential increases in the severity of precipitation due to climate change (Min et al., 2011; Intergovernmental Panel on Climate Change [IPCC], 2012).

Disruptions to installations could emanate from different locations along the electricity fuels, extraction (if applicable), transportation, generation, transmission, and distribution supply chain. DoD planners can think about power outages in terms of the frequency of events each year and the duration of individual events. An examination of power outage data from 1984 to 2006 found that blackout frequency has not decreased over time and that blackouts are more frequent in the summer and winter and during peak usage hours of the day (Hines, Apt, and Talukdar, 2009). Using DOE criteria for major disruptions in the United States, there were at least 30 electricity disruptions in 2011 that affected 250,000 or more customers, with five of these events each affecting more than 1.5 million customers (DOE, 2012a). These estimates do not include the myriad other disruptions not characterized as major events (e.g., Eaton Corporation, 2012) but that nonetheless could affect DoD capabilities. The durations of power outages can vary depending on the cause, extent of damage, and local conditions. Hurricane Sandy, which made landfall on the U.S. East Coast on October 29, 2012, disrupted power for more than 8.5 million customers—650,000 of which were still without power 11 days later. Because electricity is also required for the distribution and pumping of gasoline, nearly a quarter of gas stations in the New York metropolitan area were still unable to sell gasoline more than a week after the hurricane's impact (DOE, 2012b). During the powerful derecho storm in 2012 that resulted in the loss of power to 3.8 million customers in the U.S. Midwest and Mid-Atlantic areas, the local utility in Washington, D.C., restored service to most of its nearly 450,000 customers in five days, with service to all its customers restored in ten days (Pepco, 2012). Situations such as these illustrate the need for DoD planners to consider extended local and regional outages in energy security planning.

Many U.S. installations also rely on the commercial natural gas distribution system for natural gas used for heating, hot water, food preparation, and in some cases distributed electricity generation. Similar to the electricity supply chain, disruptions to end users could emanate from different locations along the fuel production, processing, transmission, storage, and distribution networks. There were five major U.S. natural gas supply disruptions in 2011 (DOE, 2012a). End users of natural gas can be disrupted from similar causes as electricity system disruptions, as well as from an electricity outage itself, if electricity is required for natural gas compression stations or other natural gas infrastructure. In 2011, a cold front in the Southwest resulted in the loss of natural gas production as a result of frozen water in natural gas lines that could not be repaired because of icy roads, as well as the loss of electricity to natural gas compression stations. These conditions left more than 50,000 customers in New Mexico, Arizona, and Texas without natural gas for up to one week during a period of very cold temperatures, because a labor-intensive safety process must be undertaken to restore service to each customer after an outage (Federal Energy Regulatory Commission [FERC] and North American Electric Reliability Corporation [NERC], 2011). In New Mexico, the governor declared a state of emergency and ordered National Guard Troops to assist in restoring service (Smith, 2011). Finally, just as electricity outages can affect the natural gas system, electricity producers with considerable natural gas generation assets would be affected during a natural gas shortage (FERC and NERC, 2011). Electricity was disrupted for 4.4 millions customers in the Southwest during this cold-weather event in 2011. Although most of the electricity outages were due to the cold weather itself, at least 12 percent of outages were due to the affected natural gas supply (FERC and NERC, 2011).

Although DoD can provide input and assistance to those responsible for enhancing the security of the commercial electricity and natural gas networks, it has rightly refocused efforts on domains within its control—enhancing energy security at installations through a mix of technologies and strategies. Installations can install their own electricity generation equipment, and some installations purchase electricity from utilities or companies that construct, operate, and maintain energy assets on or adjacent to installations (see, for example Environmental Protection Agency [EPA], 2009). Installation personnel provide physical security for energy assets on or adjacent to DoD installations. Yet unless these assets are designed with the capability to provide power to the installation during a wider grid blackout, they may not be operational during an emergency and will be of little to no value from an energy security perspective. Most installations currently plan for the loss of energy services by connecting emergency diesel-powered generators to specific facilities that have been deemed critical. These generators are designed to sustain basic installation functions and critical missions for 3–7 days using available on-site fuel storage (Stockton, 2011a). The Government Accountability Office (GAO) noted that DoD had not considered vulnerabilities to longer-term power outages (2009a) and that DoD's process for identifying its critical infrastructure assets needed improvement (2009b). Furthermore, the current approach can increase energy security to each isolated facility, but the lack of a systems approach to energy security could result in some critical capabilities being unavailable, the additional expense of purchasing and maintaining separate diesel generator networks, and missed opportunities to achieve synergies and potential cost savings from coupling DoD's energy security, environmental, and renewable energy goals.

In addition to providing military capabilities, installation energy services also enable DoD to support the Department of Homeland Security during disasters. Stockton (2011b) argues that DoD needs to be better prepared to provide support in the event of a "complex catastrophe"—one whose scale and destruction are significantly greater than normal disasters and can potentially result in the "cascading, region-wide failures of critical infrastructure." Social services disrupted during a complex catastrophe could include electricity, water, transportation, health, and other functions (Stockton, 2011b)—as demonstrated in the aftermath of Hurricane Sandy. Whereas regional distributed generation, distributed automation, and advanced metering could potentially sustain some critical services during extended blackouts, these investments would need to be made by others outside DoD and span traditional political and decisionmaking boundaries. Communities that choose to invest in these technologies could become islands of stability during a complex catastrophe but might also find themselves inundated by people from areas without services (Narayanan and Morgan, 2012). Energy security planning at DoD installations should consider the possibility of similar situations and recognize the need to plan for complex regional catastrophes.

To provide energy security on DoD installations in the United States, DoD needs a strategy to maintain critical capabilities and abide by what Davis (2011) called the FARness principal. Davis stressed the need for analysis to help decisionmakers identify strategies that have *flexibility* for diverse missions, *adaptiveness* in various circumstances, and the *robustness* to withstand and recover from adverse events. In this report, we present the foundation of a potential capabilities-based framework for evaluating energy security decisions on U.S.-based DoD installations.

Outline of This Report

This report is structured as follows. In Chapter Two, we introduce the concept of using DoD Joint Capability Areas (JCAs) to serve as a basis for measuring effectiveness in evaluating installation energy security strategies. In Chapter Three, we outline and describe how a capabilities-based planning approach using portfolio analysis could be performed for evaluating energy security strategies at U.S.-based installations. We discuss conclusions in Chapter Four. Finally, the appendix lists what we see as potentially the most relevant Joint Capability Areas that are enabled by access to installation energy services.

CHAPTER TWO

Using Joint Capability Areas to Inform Installation Energy Security Decisions

DoD analysts planning for installation energy security rely on the broad energy security definition established in the 2012 National Defense Authorization Act—assured access, reliable supplies, and sufficient energy to meet mission essential requirements, as described in Chapter One. The remaining challenges for decisionmakers throughout DoD are operationalizing and acting on these concepts. In this chapter, we discuss the initial task of measuring installation energy security and outline how DoD Joint Capability Areas might serve as a basis for these measures. This type of exercise can and should be undertaken for each installation individually, but a broader, DoD-wide capabilities assessment could inform decisionmakers on maintaining overall capabilities during the loss of energy services.

Defining Joint Capability Areas for Homeland Security and Emergency Response

DoD is the primary agency responsible for *homeland defense*, or the military protection of the United States from external threats and aggression. DoD also provides support to other civilian agencies (if needed) in providing *homeland security* to prevent terrorist attacks on the United States, reduce vulnerability and minimize damages from terrorism, and assist in the recovery from terrorist attacks or natural disasters.

The DoD Assistant Secretary for Homeland Security leads DoD's efforts in homeland defense and security and is also DoD's liaison on these issues to the Department of Homeland Security, the National Security Council, and the White House (Bowman, 2003). Goss (2005) argues that capabilities-based planning, rather than threat-based planning, is more appropriate for DoD homeland defense and homeland security planning, because of the multiple overlapping actors involved and the amorphous threats faced. Our discussion about installation energy security is one aspect of this broader planning challenge.

The 2001 Quadrennial Defense Review introduced the concept of shifting from a threat-based defense planning model to a capabilities-based approach. The idea focused on identifying and developing the capabilities required to successfully address the planning uncertainties of surprise, deception, and asymmetric warfare. This began a shift toward planning for "how an adversary might fight" rather than specific scenarios of who and where they are (DoD, 2001). After the publication of an independent capabilities report in 2004 (Joint Defense Capabilities Study Team, 2004), DoD divided the department's capabilities into functional categories to enable capabilities-based planning for warfighting needs (DoD, 2010b). A series of JCAs were developed and refined, with a Joint Capability Area Management Plan issued in 2010 (DoD, 2010b) and a 2011 Memorandum for the Vice Chairman of the Joint Chiefs of Staff outlining the 2010 JCA refinement (DoD, 2011a). DoD has defined nine primary JCAs: Force Support, Battlespace Awareness, Force Application, Logistics, Command and Control, Net-Centric, Protection, Building Partnerships, and Corporate Management and Support. Each primary JCA is composed of increasingly detailed second, third, fourth, or further lower-tier JCAs (full definitions are provided in DoD, 2011b). For example, under the Net-Centric JCA, the sub-JCA Information Transport is defined as "The ability to transport information and services via assured end-to-end connectivity across the [Net-Centric] environment." DoD codified the relationship between tasks, capabilities, effects, and objectives as presented in Table 2.1.

Although the current DoD JCAs are clearly written to define many capabilities required for homeland defense, their applicability in supporting homeland security functions is less clear. For example, under the Logistics primary JCA, the sub-JCA Facilities Support is defined as, "The ability to provide functional real property installation assets with utilities—energy, water, and wastewater; contract and real property management; pollution prevention; and essential services throughout natural or man-made disasters." Also under Logistics, the Emergency Services sub-JCA is defined as,

Table 2.1
Relationship Between Tasks, Capabilities, Effects, and Objectives

DoD Planning Term	DoD Planning Definition and Relationship
Tasks	An action or activity assigned to an individual or organization to provide capability.
Capability	The ability to achieve a desired effect under specified standards and conditions through a combination of means and ways across the Doctrine, Organization, Training, Materiel, Leadership and Education, Personnel, and Facilities (DOTMLPF) to perform a set of tasks to execute a specified course of action. Capabilities result from a combination of tasks.
JCA	Collections of like DoD capabilities functionally grouped to support capability analysis, strategy development, investment decisionmaking, capability portfolio management, and capabilities-based force development and operational planning.
Effect	A change in condition, behavior, or degree of freedom. Capabilities are applied to create desired effects.
Objective	A desired end derived from guidance. Objectives are achieved by creating desired effects.

SOURCE: Adapted from DoD, 2010b.

"The ability to protect and rescue people, facilities, aircrews, aircraft and other assets from loss due to accident or disaster" (DoD, 2011b).

Yet, we posit that the existing JCA framework can be used as the basis for capabilities-based metrics for installation energy security analyses. The 2008 National Defense Authorization Act requires that DoD, in consultation with the Department of Homeland Security, "determine the military-unique capabilities needed to be provided by the Department of Defense to support civil authorities in an incident of national significance or a catastrophic incident" (P.L. 110-181, 2008). The nature and extent of DoD capabilities required for civil support that are enabled by installation energy could be defined during a fully developed analysis and incorporated into individual metrics to evaluate energy security options.

This report is intended to stimulate discussion around the issue of how installation energy services contribute to homeland defense and homeland security and how DoD can plan to maintain these capabilities. It is also meant to demonstrate how capabilities-based planning could be used to analyze energy security strategies. For this demonstration, we draw upon the existing JCAs, recognizing the limitations of this approach and the potential need for continued refinement of the JCAs for applications across homeland defense and homeland security.

Illustrative DoD Capabilities for Energy Security

As U.S. DoD installations are integral components of maintaining DoD capabilities, we propose using JCAs to form the foundations of metrics to evaluate installation energy security decisions. In essence, we can ask if existing or proposed installation energy security strategies enhance DoD capabilities and evaluate strategy cost-effectiveness. Using a common framework and systems approach, DoD can link and evaluate how energy security tasks and strategies affect DoD-wide JCAs during a loss of installation energy services that may vary in complexity, space, time, and mode. As we will discuss in Chapter Three, analysts using this framework would evaluate how energy security options perform across multiple dimensions of JCAs.

Analyzing installation energy security decisions using JCAs would begin with identification of the most critical JCAs and sub-JCAs that could be affected by a loss of installation energy services. Overall, both JCAs and sub-JCAs will be affected by the loss of installation energy services differently. For instance, loss of power to an airfield will affect aircraft operations, affecting the Logistics primary JCA, but loss of energy services would have a lesser impact on DoD's ability to negotiate partnership agreements with domestic and foreign institutions, affecting the Building Partnerships primary JCA. In the appendix, we present the potentially most relevant second-, third-, and fourth-level JCAs that are enabled by access to installation energy services. We use these selections as a starting point for discussion and prioritization; a full analysis would examine each JCA in depth to gauge the effects of a loss of energy services.

Expert interviews, site visits, modeling, and analysis of completed critical infrastructure planning could inform a detailed analysis of the JCAs.

After the most critical JCAs and sub-JCAs have been identified, the capabilities used for analysis should be defined and sharpened so that they are most relevant for decisionmakers (Davis, Shaver, and Beck, 2008). Using capability categories allows for the crosscutting objectives of each JCA to be measured within a common category. For example, the Logistics JCA includes such diverse tasks as moving equipment, feeding installation personnel, and providing installation law enforcement. Measures of effectiveness in evaluating strategies to maintain a Logistics JCA would necessarily encompass multiple capabilities. To measure effectiveness across this and the other broad JCAs, we first need to identify specific functions required by the JCA and then identify well-defined measures. Looking across the diverse JCAs and sub-JCAs, we observed several primary DoD responsibilities that could serve as top-level, plain English functions for installation energy security planning. These include the provision of training, information, materiel, care, and security, as listed in Table 2.2. We stress that these defined functions are an initial suggestion and should be refined under a fully developed analysis for installation energy security. To assess effectiveness across each function under various energy security strategies, several analytic measures of effectiveness would serve as metrics and underlie the assessment of each function. For example, the Provision of Care function would include such measures of effectiveness as the percentage of hospital beds capable of handling patients, the percentage of active-duty personnel housing capable of housing personnel, and, potentially, the capabilities associated with incidents outside the installation, such as the number of triaged patients per hour in the aftermath of a natural disaster.

Table 2.2
Mapping of DoD Joint Capability Areas to Measures of Effectiveness in Evaluating Installation Energy Security Strategies

DoD JCA	Provide Training	Provide Information	Provide Materiel	Provide Care	Provide Security
1. Force Support	X			X	
2. Battlespace Awareness		X			
3. Force Application		X	X		X
4. Logistics		X	X	X	X
5. Command and Control		X			
6. Net-Centric		X			X
7. Protection		X			X
8. Building Partnerships		X			
9. Corporate Management and Support		X	X		

Developing a Capabilities-Based Approach for Evaluating Energy Security Decisions

DoD is currently engaged in evaluating installation energy security strategies, an undertaking that will need to incorporate technological, economic, and operational uncertainties. In Chapter Two, we proposed adapting DoD Joint Capabilities Areas as the foundation for measuring energy security at installations. In this chapter, we build on this concept and describe how a capabilities-based planning approach for evaluating energy security strategies at U.S.-based installations could be performed.

Capabilities-based planning (CBP) means different things to different people and some aspects of its implementation in DoD have been appropriately controversial. Previous RAND work has defined CBP as planning under uncertainty to provide capabilities for a wide range of challenges (including diverse circumstances) while working within economic constraints (Davis, 2002; Davis, Shaver, and Beck, 2008). This means making choices in allocating limited resources to be in the best position to deal with future demands, which cannot be perfectly anticipated or defined in advance. In this context, *capabilities* includes the broad meaning associated with "ability and wherewithal," not just "assured ability to do a very specific task in a very specific set of circumstances." CBP is inherently about making choices; it is the opposite of a blank-check approach. Finally, despite confusion on the matter a decade ago, CBP also includes using concrete scenarios to test the effectiveness of options. Such scenarios should be chosen analytically so as to be good test cases of broad capability, with no illusions about their being meaningful "best estimates." This interpretation of CBP is quite consistent with the most recent Quadrennial Defense Review, and CBP is now a component of DoD decisionmaking (DoD, 2011b). Previous RAND analysis described enabling analytic tools for a CBP approach (e.g., Davis, 2002; Davis, Shaver, and Beck, 2008; Davis et al., 2007; Davis and Dreyer, 2009).

The first step in a capabilities-based analysis is to define the capabilities needed, as discussed in Chapter Two. The remaining steps are characterizing the broad analytical scenario space and developing a spanning set of scenarios, generating investment building blocks and screening composite options, evaluating options in a portfolio analysis, and, finally, characterizing shortfalls and iterating for improvement (Davis, Shaver, and Beck, 2008). We use these steps to illustrate how a capabilities-based approach could be used for evaluating installation energy security decisions.

Characterizing the Broad Range of Future Challenges

In planning, it is essential to consider a broad range of future challenges—what is often referred to as a broad "scenario space," with the word *scenario* referring not just to a generic category, such as a natural disaster, but to a specific example with all of the details that fully define it. The *parametric uncertainties* inherent in the input factors that influence results in a specific scenario might include such details as location or weather conditions or the strategies and behaviors used by adversaries during a terrorist attack. The *structural uncertainties* of how a model represents and values the relationship between inputs also expand the range of responses to a specific scenario (Davis, 2012). An infinite number of such scenarios exist, so—after thinking about the broad scenario space—planners need a smaller set of test cases with which to work. These tests can only be illustrative but should be chosen analytically[1] to stress the options under consideration in all the important dimensions. The resulting set of test cases has been called a "spanning set," to suggest that an option that does well for all of the test cases should be able to do well in a real world case, even though that would most likely be different from any of the test cases (Davis, Shaver, and Beck, 2008).

An analytic scenario space for evaluating installation energy security strategies needs to include the considerable uncertainties and aspects involved. We begin by brainstorming some of the key dimensions associated with a loss of installation energy services and tie specific cases or examples to these dimensions, as shown in Table 3.1. The loss of services could result from a natural disaster, an accident, equipment failure, intentional disruption, or other event—or a series of these events. What mat-

Table 3.1
Key Dimensions, Cases, and Examples Associated with a Loss of Installation Energy Services

Dimension	Case or Example
Mode of loss of energy service	Loss of service from commercial electrical or natural gas grid, loss of petroleum resupply, loss of installation energy assets
Broad scenario class	Simple natural disaster, complex catastrophe, deliberate attack by determined adversary
Geographic location affected	Part of an installation, entire installation, surrounding city, surrounding region, the continental United States (CONUS)
Preparedness	Warning time, in-place capabilities for resiliency, response and repair, installation load management plan in place
Availability of enabling response resources	Water available, food available, response and repair capabilities available
Assumptions and models used to evaluate options	Amount of energy actually needed or used in specific detailed scenarios and how inputs are valued

---

[1] Several analytic methods exist for scenario discovery under uncertainty. See, for example, Davis, Bankes, and Egner, 2007; Groves and Lempert, 2007; Lempert, Bryant, and Bankes, 2008; Bryant and Lempert, 2010.

ters for strategic decisionmakers evaluating energy security strategies are the several dimensions regarding the availability of energy services, the conditions under which these losses occur, and the circumstances affecting a potential response. Can a response begin immediately, or does the risk of additional damage from additional coordinated attacks (or, for that matter, a prolonged severe storm) affect response choices? What are baseline levels of emergency power provision at existing installations, and how have these been affected? What assets can be leveraged or transported from other DoD installations? Brainstorming the potential pathways and context for disruption is important to allow unexpected dimensions to emerge.

In a fully developed analytic scenario space, analysts would identify dimensions that would expose the potential weak points in a response and the types of shocks that would occur and would think about what would enhance adaptiveness. The illustrative events that we identify will typically be individually unlikely, but the likelihood of at least one of them occurring is quite high. Of course, some events will occur that were neither anticipated nor very likely in any of the examples. They represent events that officials often declare "no one could have anticipated"—this underscores the importance of both rigorous thinking about scenarios, as well as planning for adaptiveness after unforeseen events.

Installations predominately derive energy services through the use of energy commodities produced by others externally and delivered to the installation via public infrastructure. These include commercial electricity fuel extraction (if applicable), production, transmission, and distribution systems; commercial natural gas extraction, processing, storage, transmission, and distribution systems; and the global and national markets for crude oil extraction, transportation, refining, storage, and distribution systems. Each node in these supply chains represents an opportunity for accidental or intentional disruption, which could potentially affect the installation's ability to receive these commodities as expected. As described in Chapter One, energy service disruption can also be cascading—the loss of electricity disabling local petroleum product dispensing is a notable experience from Hurricane Sandy. Electricity, natural gas, and petroleum generally provide different types of energy services on an installation, although some near-term and mid-term fungibility and transformations exist. Electricity provides lighting, communications, computing, air-conditioning and ventilation, health care diagnostics, refrigeration, electric mobility, and many other services. Natural gas provides space and water heating, food preparation, manufacturing process heat, and other services but can also be transformed into electricity via a generator, engine, or fuel cell. Finally, petroleum is refined into distinct products that can provide tactical, nontactical, and training mobility, but these refined products can also be transformed into electricity or heat via a generator, engine, boiler, or fuel cell. Petroleum products enable some baseline level of emergency power generation for critical capabilities at all current installations. In addition, some current installations and future capabilities include the distributed generation of electricity or heat within the installation with renewable or nuclear fuels. As such, the distributed generating assets themselves, as well as their enabling infrastructure, remain vulnerable to disruption.

Using these energy pathways and the key dimensions of the broad scenario space relevant to installation energy security, we can create a spanning set of test cases. Davis, Shaver, and Beck (2008) define a spanning set of scenarios as "a set of test scenarios chosen so that if alternative proposed systems are tested against them, the systems will be 'stressed' in all the appropriate ways. Systems that do well across the test cases should do well in the situations that arise in the real world." For DoD installations in the United States, we define the four broad scenario classes for our spanning set of test cases in Table 3.2.

The robustness of any strategy to respond to any individual scenario or combination of these critically depends on the characteristics and circumstances of the individual scenarios. Planning for resiliency against the loss of installation energy services for one hour versus several months necessarily requires different approaches. Is the electricity outage due to downed trees from a summer thunderstorm or from an ice storm that also degrades roadway functionality? What response capabilities are available during simultaneous cyber attacks on the electricity and natural gas systems by a determined adversary? What kinds of services will DoD installations need to provide to the local community during a complex catastrophe? Hence, the four test cases should be analytically stressed along the dimensions of complexity, scale, time, preparedness, and response resources.

We outline the relevant parameters for our spanning set of scenarios in Table 3.3. Analysts would use parameters such as these to test the performance of various strategies across a spanning set of scenarios.

We illustrate our four test cases in Figure 3.1. The key dimensions of each test case can be parametrically explored with a capabilities-based analysis. Probabilities of each of these dimensions, and for each test case, could be generated through histori-

Table 3.2
Broad Scenario Class and Energy Services Disrupted

Broad Scenario Class	Energy Services Disrupted
Loss or reduction in electricity from the commercial grid	C4ISR, lighting, space/water heating, air conditioning, ventilation, water/sewer, health care diagnostics/provision, food preparation/storage, manufacturing, fuel distribution
Loss or reduction in natural gas service from the commercial distribution system	Space/water heating, backup or primary electricity generation, food preparation, manufacturing
Loss or reduction in petroleum-based fuel supplied to installations	Backup or primary electricity generation, heating, tactical mobility, nontactical mobility
Loss or reduction in energy assets within an installation	Distribution and management of electricity, natural gas, and steam; on-site provision of electricity and heat via renewable and conventional fuels; energy storage and controls

Table 3.3
Parameters for Evaluating Energy Security Strategies Across a Broad Scenario Space

Complexity	Scale	Disruption Time	Preparedness	Response Resource
Simple accident or natural disaster	Partial installation	1 minute to 180 days	0 to 7 days of warning	Water/food availability
	Entire installation			
Complex catastrophe			Load management plan availability	Response/repair capabilities
	Surrounding community			
Determined adversary				
	Entire Command region		Available emergency assets	

Figure 3.1
Four Spanning Test Cases Illustrating Pathways for Installation Energy Services Disruption

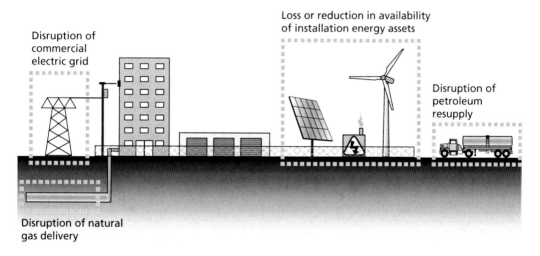

NOTE: Illustrative and not to scale.
RAND RR162-3.1

cal analysis, expert elicitation, and other methods. For example, petroleum products can be resupplied to an installation via multiple pathways and transportation modes, and the probability of a CONUS-based installation being physically unable to receive logistical petroleum shipments for an extended time period is low. However, during a regional complex catastrophe, the commercial refueling infrastructure relying on electricity to operate could be severely degraded and would face insatiable demand (Stockton, 2011b; Narayanan and Morgan, 2012), as experienced during Hurricane Sandy. Analysts and decisionmakers using a capabilities-based approach can look across and within the ability of options to respond to test cases under a range of event probabilities and from different perspectives.

It is also important to define the potential concept of operations (CONOPS) and critical components for evaluating options. For installation energy security strategies, these might include ensuring that solutions to technical and personnel challenges exist before declaring that a specific option has operational capability (Davis, Shaver, and Beck, 2008). An example might be the existence of trained energy managers on the installation who can execute an emergency energy plan or who have the requisite clearances to participate in the planning process (Lachman et al., 2011).

Identifying Investment Building Blocks for Candidate Options

Maintaining capabilities during the loss of energy services can be accomplished with various energy assets and strategies. Generally, the individuals considering options to meet a specific capability develop the options based on their organization's past efforts, knowledge, and interests. Developing strategies this way limits the opportunities for synergies across capability areas, the incorporation of uncertainty parameters on costs and capabilities, and the ability for decisionmakers to evaluate tradeoffs and risks. These challenges can be mitigated through a systematic approach to option generation where analysts screen and identify composite options for evaluation (Davis, Shaver, and Beck, 2008; Davis et al., 2007).

The process for generating composite options from building blocks is detailed in Davis et al. (2007). Analysts begin by identifying individual building blocks and their cost and performance characteristics. Then they construct the set of all possible composite options, which consist of all possible combinations of the building blocks. The analysts then screen the large set of possible options to eliminate those that do not meet effectiveness or cost thresholds, as well as those that are clearly inferior to other options (Davis, Shaver, and Beck, 2008). Instead of retaining composite options only on the efficient (Pareto-optimal) frontier between effectiveness and cost, analysts using this method also keep options near the efficient frontier, as these options may emerge as dominant under more rigorous portfolio analysis. Finally, the analysts construct a limited set of composite options that are at or near the efficient frontier for at least one of the test cases. The result is to generate a richer set of options than if "requirements" had been narrowly defined early and the correspondingly "optimal" (but actually suboptimal) options had been identified for each level of cost. These enhanced details are important when planning under uncertainty.

For strategies to maintain capabilities during a loss of energy services, we list sample building blocks and the motivation for including these in Table 3.4. We characterized the sample building blocks into three separate groups: (1) efficiency, operations, and information investments; (2) energy generation, distribution, and control capital equipment; and (3) fuels, energy storage, and enabling equipment capital purchases. The performance of each building block would depend on assumptions about local

Table 3.4
Sample Building Blocks and Motivation for Energy Security

DoD Investment Building Block	Motivation for Energy Security
Efficiency, Operations, and Information Investments	
Investment in installation energy efficiency assets or information campaign	Reduces primary and emergency installation electricity and heating needs
Electronically and physically secure information technology; Supervisory Control and Data Acquisition systems and other electronic power control systems	Reduces likelihood of electronic or physical attack on enabling infrastructure systems
Develop and implement emergency load management plan	Provides for essential operations and reduced energy loads during emergency conditions
Command-level situational assessment capabilities related to energy and outages	Information across command region will dictate response
Rapid adaptive planning capabilities and agreements for cooperation with the Department of Homeland Security and other agencies	Defines potential DoD responsibilities, which inform needed capabilities
Energy Generation, Distribution, and Control Capital Equipment	
Diesel generators	Generate emergency electricity
Natural gas distributed generation	Generates primary or emergency electricity, heat, and cooling
Solar photovoltaic (PV) or concentrated solar power distributed generation	Generates primary or emergency electricity and heat
Wind distributed generation	Generates primary or emergency electricity
Waste-to-energy distributed generation	Generates primary or emergency electricity and heat
Fuel cell	Generates primary or emergency electricity and heat
Small modular nuclear reactor	Generates primary or emergency electricity and heat
Full or partial installation microgrid coupled with secure connections to commercial grid and installation assets	Enables control and electricity distribution from local generation assets
Fuels, Energy Storage, and Enabling Equipment Capital Purchases	
Acquire additional supply of stored diesel or other liquid drop-in fuel, which may involve construction of storage facilities	Permits operation of diesel generation for defined period
Construct natural gas storage facilities and acquire supply of stored natural gas	Permits operation of natural gas generation for defined period
Acquire energy storage assets, either standalone or through leveraging nontactical vehicle fleet	Stores electricity produced by commercial electric grid or distributed generation and provides primary or emergency electricity
Acquire spare generation, controls, distribution, and other enabling equipment	Provides for redundancies and enhances ability to repair damages

resources, and the technical and economic characteristics of each building block, as well as the presence and synergies of other building blocks (e.g., microgrid availability and renewable electricity). An analyst generating composite options with energy assets and strategies would necessarily need to incorporate the uncertainty of current and future costs and performance for each building block to fully represent the range of possibilities involved with each composite option. A set of energy security building blocks would represent a composite option for analysis. The potential to deliver additional energy assets from unaffected installations would be considered as part of the command-level situation assessment, and existing assets providing emergency energy services at each installation would be considered within the context of emergency load management plans. The demand reductions enabled by emergency load management plans would reduce the amount of energy assets needed, and a thorough analysis of installation capabilities under considerable demand reductions is an essential step in installation energy security analyses.[2]

Assessing Effectiveness with a Capabilities-Based Portfolio Analysis

Once analysts generate composite options of technologies and strategies, they can conduct capabilities-based planning under uncertainty using portfolio analysis. This allows the effectiveness of each option to be tested across the broad scenario space. RAND previously developed the Portfolio Analysis Tool (PAT) to assist in capabilities-based portfolio analysis that addresses both uncertainty and differences in perspective. A portfolio approach and tools such as PAT allow analysts to organize and evaluate complex problems with multiple objectives and multiple solutions depending on the importance weighting of each objective (or on nonlinear evaluation of overall value, which goes beyond weighting). Various risks of each option, such as technological, strategic, political, programmatic, operational, and others, can be integrated and aggregated across the analysis (Davis, Shaver, and Beck, 2008; Davis and Dreyer, 2009).

For evaluating installation energy security strategies, we propose testing the effectiveness of a set of generated technology and strategy options of providing capabilities across our defined broad scenario space outlined above. The capabilities-based metrics would align with each of our top-level measures of effectiveness outlined in Chapter Two: the provision of training, information, materiel, care, and security. Each top-level measure of effectiveness would be composed of a set of individual metrics that would be measurable and comparable across options. That is, each option would be evaluated on how well it performs at providing training, information, materiel, care, and security under broad scenarios testing the loss of installation energy services. As discussed, the

---

[2] For example, after the 2011 earthquake in Japan, a focused public energy efficiency campaign reduced the effects of electricity supply shortages.

parameters of each scenario—complexity, scale, disruption time, preparedness, and response resources—test the strength of different options across the broad range of potential situations. Therefore, the effectiveness of each option is summarized for the top-level measures but is evaluated by aggregating how the options perform across the breadth of individual metrics and the depth of each parameter. Davis, Shaver, and Beck (2008) call this process "drilling down" to subsequent levels, which can help identify particular areas where capabilities are deficient. Each drill-down level can also provide additional information to decisionmakers regarding why a particular option is rated higher than another option (Davis, Shaver, and Beck, 2008; Davis and Dreyer, 2009). This process is depicted in Figure 3.2.

In using PAT for evaluating installation energy security strategies, the summary chart would be a stoplight chart that allows analysts to schematically compare how the options perform across measures of effectiveness for installation energy security. As illustrated in Table 3.5, top-level measures of effectiveness would comprise columns in the summary scorecard, whereas the composite options developed to address energy security would comprise the rows. In the summary chart, each color represents an aggregated measure of broad capabilities maintained by each option, enabling a decisionmaker to quickly scan across options to understand how that option contributes to capabilities for energy security. Users can change the way underlying calculations

Figure 3.2
Hierarchy of Detail in RAND's Portfolio Analysis Tool

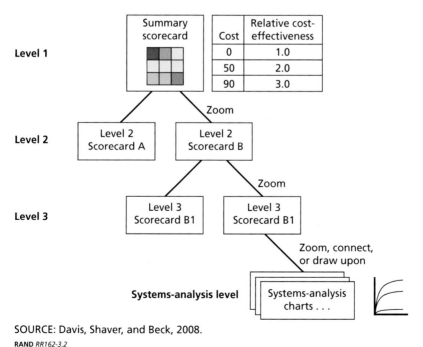

SOURCE: Davis, Shaver, and Beck, 2008.
RAND RR162-3.2

Table 3.5
Illustrative Example of Top-Level Energy Security Evaluation Across Multiple Options

Option	Provide Training	Provide Information	Provide Materiel	Provide Care	Provide Security
A					
B					
C					
D					

are made with a dashboard of options and assumptions. Additionally, we show in Figure 3.2 numerical tables illustrating how each option's overall cost and cost-effectiveness would accompany the summary scorecard (Davis, Shaver, and Beck, 2008; Davis and Dreyer, 2009).

We note that in conducting a capabilities-based approach for installation energy security strategies, one could of course arrange the analytical space differently than as we have proposed. Instead of having the top-level summary chart organized by the JCA functional classes, the summary chart could evaluate composite options across each of the scenario classes. The second level could then be the key dimensions of each of theses scenarios (as described in Table 3.2), and options could then ultimately be evaluated across measures that span the scenario classes. We chose to propose a top-level chart organized by JCA classes, because we posit that decisionmakers for this application would be better served by evaluating the ability of a composite option to provide a set of capabilities (providing materiel, care, etc.) across a range of energy service disruption scenarios than by determining how well an option performed during the loss of electricity services. Additionally, because of the interdependencies between energy service disruption scenarios (e.g., loss of electricity *and* natural gas), arranging the top-level summary chart by JCA classes allows for performance comparison across all scenarios in the top level. Ultimately, the method chosen would depend on the questions being asked and the value of specific analytical presentation styles to decisionmakers.

We illustrate one way the analysis could be conducted with an example. As we discussed, we propose evaluating the performance of several installation energy security composite options on five top-level measures of effectiveness (the provision of training, information, materiel, care, and security) over a spanning set of broad scenario classes (loss of electricity, natural gas, petroleum, and on-site assets) and their parameters (complexity, scale, disruption time, preparedness, and response resources). Each top-level measure of effectiveness is composed of several individual metrics derived from

DoD's JCAs.[3] A fully developed analysis would carefully derive and define the appropriate set of individual metrics. For our example, we define one individual metric—the common logistics capability of airfield management, which ensures safe and timely aircraft operations on installations. This could be one of the individual metrics composing our Providing Materiel top-level measure.

To efficiently and effectively manage aircraft takeoffs and landings, airfields require power for tower communications, airfield lighting, ground-based instrument approach systems, and, potentially, other functions. This power is provided by the electricity grid and is backed up with emergency diesel generators. The loss of energy services at an installation would therefore degrade airfield operations. Under normal operating conditions, a specific installation can provide a range of takeoffs and landings per day. A capabilities-based metric could be the number of aircraft operations (takeoffs and landings) per 24 hours, provided by the composite options across the spanning set of test cases and parameters. Under each composite option, all of which are composed of technology and strategy building blocks, the number of daily aircraft operations possible would be evaluated across the spanning set of test cases. Some simple example composite options could feature enhancing existing diesel fuel stockpiles, solar PV generation coupled with energy storage, or a natural gas/diesel dual-fuel generator connected to serve airfield power needs—although actual composite options could use combinations of building blocks. Evaluating the performance of these options to maintain airfield operations, combined with the myriad other individual metrics, would form the Level 1 results shown on the summary scorecard. The results would depict how individual composite options performed across all the spanning test cases.

The capabilities framework we outlined previously (based on Davis, Shaver, and Beck, 2008) uses drill-down levels for each parameter of a spanning test case. Continuing our example, the individual metric of maintaining airfield operations is evaluated for each of the spanning test cases and across each parameter. For instance, an option relying heavily on existing diesel generators may perform well at providing backup power for airfield operations for short durations but may be less competitive for longer durations of outages than an option relying more on renewables, microgrids, and energy storage. The Level 2 measures would test how a specific option performed under the loss of electricity, natural gas, petroleum, or installation assets. For each spanning test case (such as the loss of installation electricity), each option is evaluated for its effectiveness in maintaining airfield operations depending on the complexity the disaster, the geographic scale of the outage, how long the duration lasts, how prepared the installation is, and what resources are available to respond. Each parameter would represent subsequent levels of measures. One analysis pathway is illustrated in Figure 3.3, and a sample scorecard for several levels is shown in

---

[3] From the list of DoD JCAs, we list some primary examples and priorities to focus on for developing capabilities-based metrics in the appendix.

Figure 3.3
Illustrative Example of Test Cases and Dimensions to Measure Effectiveness of Installation Energy Security Strategies

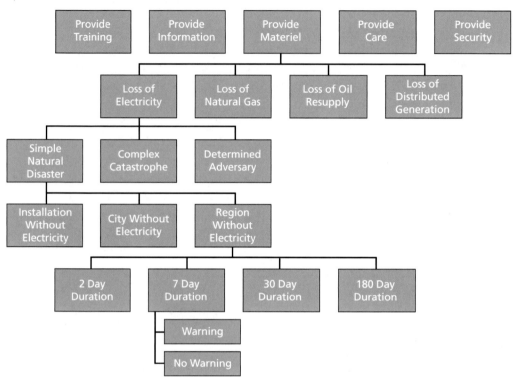

Figure 3.4. The performance of composite options for airfield operations would be evaluated for the loss of electricity during a simple natural disaster or accident, for a complex catastrophe that would require a coordinated emergency response, and for an attack by a determined adversary. For each measure, the performance of composite options for airfield operations would be evaluated for when an installation is without electricity, the city is without electricity, or the entire region is without electricity. This would continue drilling down to subsequent levels to expose option deficiencies across parameters. For each composite option under evaluation, the effectiveness score for each level is the aggregated evaluated effectiveness score for the underlying hierarchy, as illustrated in Figure 3.4. This analytic evaluation process, conducted in an open, documented, computer programming environment, would be repeated across options, individual metrics, and parameters for the broad scenario space.

It is now apparent that, with multiple measures of effectiveness and several levels of test case dimensions, the summary results of capabilities-based analyses depend highly on the weighting and aggregation rules for lower-level assessments. These rules should reflect the priorities of the overall assessment, obtained through iterative dis-

Figure 3.4:
Ilustrative Example of the Drill-Down Process and Measures as an
Aggregation of Performance for Each Composite Option

Option	Provide Training	Provide Information	Provide Materiel	Provide Care	Provide Security
A					
B					
C					
D					

Option	Loss of Electricity	Loss of Natural Gas	Loss of Oil Resupply	Loss of Distributed Generation
A				
B				
C				
D				

Option	Simple Natural Disaster	Complex Catastrophe	Determined Adversary
A			
B			
C			
D			

Option	Installation Without Electricity	City Without Electricity	Region Without Electricity
A			
B			
C			
D			

cussion of the analysis with DoD leadership. Defining how the results are aggregated requires judgment on the part of the analyst, and this judgment should be transparent for other stakeholders. Several methods of effectiveness aggregation exist, and Davis and Dreyer (2009) discuss common methods and their benefits and challenges. The first is the *Thresholds* method, which emphasizes critical systems components and characterizes a level's measure as failing if any of its measures in the subsequent hierarchy are below a defined threshold. For example, decisionmakers may arrive at the conclusion that an installation needs the capability to achieve at least 50 percent of normal aircraft operations during the loss of energy services, and any option not achieving this metric (at any level of the analysis) is assigned a failing score. After the threshold value is met, scores for that metric increase linearly until a defined goal is reached. For example, analysts could define 90 percent of airfield operations as better than 70 percent but assign no additional value for exceeding 90 percent. Another method is *Weakest Link*, which assesses overall effectiveness as the lowest of the measured scores. A third method is *Weak Thresholds*, where a value not exceeding a threshold is zero, but the overall score in the measure is weighted, rather than zero as in the *Threshold* method. For example, in an option where airfield operations are below 50 percent of the normal rate during a power outage of 180 days, its overall effectiveness for duration would be a weighted value on how this option performed for all power outage durations. Davis and Dryer (2009) also present several less common aggregation methods, but they also highlight the importance of flexibility for analysts to determine custom aggregation goals.

Evaluating Outputs and Improving Inputs

Through the analysis, each option will contain cost and effectiveness estimations for each measure considered. Yet, the capital-intensive investments and strategies undertaken for energy security will have a defined service life and provide distinct risk-reduction capabilities over this period. Analysts using a capabilities-based portfolio and defined time-horizons approach can construct a cost-effectiveness landscape of the composite options over a defined time period (Davis and Dreyer, 2009). This cost-effectiveness landscape is essentially the supply-curve of costs versus a specific method of measuring effectiveness (e.g., *Thresholds*). Comparing across composite options on a cost-effectiveness landscape, decisionmakers can understand both the costs of reaching a specific level of effectiveness and the marginal cost for achieving the next big increment of capability (Davis and Dreyer, 2009). In continuing our example of maintaining aircraft operations during a loss of energy services, a particular option might provide maintaining 70 percent of existing aircraft operations for a given cost, with the next option maintaining 90 percent of operations but at twice the cost of the first option. Decisionmakers can use information like this to judge the marginal value of increased capabilities and make tradeoffs between cost and capabilities both within and across the options for installation energy security.

We note that the perspective of the decisionmaker can substantially influence the cost-effectiveness landscape produced in the types of portfolio analyses we described. Different perspectives about time horizons, underpinned by how decisionmakers discount future costs and benefits, can change the shape of the cost effectiveness curves. RAND's previous portfolio analysis work has stressed the value of showing the cost-effectiveness landscapes for different perspectives, as illustrated in Figure 3.5 (Davis, Shaver, and Beck, 2008; Davis and Dreyer, 2009). A more challenging situation is when decisionmakers have different beliefs about how objectives have been evaluated in the analysis (see, for example, Davis et al., 2008). These situations require tailored aggregation rules so that the cost-effectiveness landscape can vary across the perspectives of the decisionmaker.

The analytical hierarchy of measures of effectiveness across test cases and dimensions allows decisionmakers to drill down and understand the drivers for shortfalls. These shortfalls could result from technical (e.g., battery charging/discharging rates), economic (e.g., capital costs for PV modules), resource availability (e.g., average wind speeds at an installation), or other reasons. DoD decisionmakers can then use the sensitivity analysis ranges established in the estimates to understand how underlying variables can affect outcomes. For many applications, the thresholds revealed by these sensitivity analyses could inform technology cost and performance targets and decisionmaking across the DoD research, development, test, and evaluation (RDT&E) enterprise. For example, an estimated cost target for energy storage on a per kilowatt-hour of energy basis that would enable greater cost-effectiveness of maintaining DoD

Figure 3.5
Effectiveness Versus Cost Landscape, by Perspective

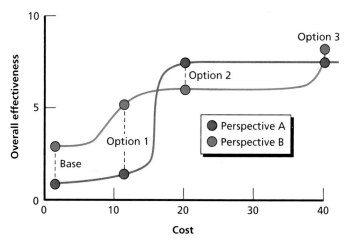

capabilities could emerge from a capabilities-based portfolio analysis and inform RDT&E portfolio managers. Thus, opportunities can be identified with a capabilities-based analysis. Finally, these analyses can help analysts identify surplus capabilities on installations, such as multiple, oversized diesel generators that are not integrated into microgrids. DoD could consider reducing the identified surpluses in one capability area to save resources for capability enhancements in other areas. The capabilities-based analysis is enhanced through an iterative process that rebalances to address shortfalls and surpluses and reevaluates capabilities and synergies after this adjustment.

DoD installations contribute to homeland defense and homeland security through providing capabilities both individually and collectively across the DoD enterprise. Installation energy security analyses and options to enhance security should therefore progress beyond the installation to the enterprise level. Just as shortfalls and surpluses are revealed on individual installations with a capabilities-based analysis, examining energy security and risks from an enterprise-level perspective can also identify such shortfalls and surpluses in capabilities. This process can reveal systemic risks under energy service loss scenarios and identify options for redundancy and resiliency for the capabilities provided by individual installations.

CHAPTER FOUR

Conclusions

U.S. installations will face rising requirements for energy security going forward, as installations increasingly directly support military operations from inside the United States. In addition to providing military capabilities, installation energy services also enable DoD to support the Department of Homeland Security's response to disasters and terrorist attacks. Our intention with this think piece is to stimulate a discussion of how DoD installation energy services contribute to homeland defense and homeland security, how DoD can evaluate choices to maintain adequate energy services to critical missions located in the United States, and how DoD can make tradeoff decisions to maintain these capabilities during disruptions.

Most DoD installations currently plan for emergency diesel-powered generators to sustain basic installation functions and critical missions for several days using available on-site fuel storage. This approach can increase energy security to each isolated facility, but the lack of a systems approach to energy security could result in some critical capabilities being unavailable, the additional expense of purchasing and maintaining separate diesel-generator networks, and missed opportunities to achieve synergies and potential cost savings from coupling DoD's energy security, environmental, and renewable energy goals.

Currently, the notion of enhancing energy security on DoD installations is challenging to plan for, execute, and evaluate. Secure for how long? Under what conditions? At what cost? And most importantly, for what reasons? We view the underlying analytical question for energy security as, "What critical capabilities do U.S. installations provide, and how can DoD maintain these capabilities during an energy services disruption in the most cost-effective manner?" We believe that installation energy security analysis is best conducted via a systems approach that incorporates technological, economic, and operational uncertainties. Capabilities-based planning (if done right) is planning under uncertainty that provides capabilities for a wide range of challenges within economic constraints. Although capabilities-based planning is now a component of DoD decisionmaking, we propose that it could be extended and used for energy security planning.

The first step in such an exercise is to define the capabilities for analysis. We propose using DoD's Joint Capability Areas as the basis for establishing metrics to evaluate installation energy security strategies. The next step is to analytically compose the broad range of future challenges for installation energy security and define a spanning set of test cases used to evaluate energy security strategies. We propose that the spanning set of test cases be (1) a loss or reduction of electricity from the commercial grid, (2) a loss or reduction of natural gas from the commercial distribution system, (3) a disruption of petroleum resupply to an installation, and (4) the loss or availability reduction of energy assets within an installation. These four test cases should be analytically expanded along the dimensions of complexity, scale, disruption time, preparedness, and response resources.

To enhance energy security at installations, DoD will use a mix of technologies (such as solar PV generation and energy storage) and strategies (such as developing emergency electricity load management plans). The next step in a capabilities-based analysis is to define composite options formed by several technology and strategy building blocks. Then the performance and effectiveness can be calculated of several composite options over the four test cases and their dimensions using portfolio analysis. The analytical hierarchy of measures of effectiveness across test cases and dimensions allows decisionmakers to drill down and understand the drivers for shortfalls. These analyses can help analysts identify surplus capabilities, then DoD could consider reducing identified surpluses in one capability area to save resources for capability enhancements in other areas. Over a range of decisionmaker perspectives, the use of cost-effectiveness landscapes generated from capabilities-based portfolio analyses can assist DoD in making choices about installation energy security strategies.

The examples described in this report demonstrate how capabilities-based planning could be used to inform choices about the adoption of technologies and practices to enhance energy security. In doing so, the report demonstrates the steps involved in this analysis and the types of data needed. The process is analytically intensive, yet it can reveal weaknesses and vulnerabilities in existing and proposed strategies to maintain installation energy services that affect homeland defense and homeland security. We have discussed our approach with some examples and detail, but these should be viewed as scoping suggestions—a fully developed DoD planning framework would incorporate mission context and relevant current issues.

Relevant Joint Capability Areas Served by Installation Energy Services

The current nine top-level JCAs are listed in Table 2.1, with the full list detailed in DoD (2011b).[1] In this appendix, we list potentially the most relevant second, third, and fourth-level JCAs that are enabled by access to installation energy services. Measuring DoD's ability to maintain these capabilities under the test cases we outlined in this report can form the basis of metrics for energy security. Although each of the nine primary JCAs are listed below, the relevant third- and fourth-tier JCAs provided here are not an exhaustive list of capabilities affected by access to installation energy services. Rather, the lower-tier capabilities are provided as primary examples and priorities to focus on while developing capabilities-based metrics.

Relevant JCAs Associated with Force Support

1. Force Support—The ability to establish, develop, maintain, and manage a mission-ready total force.

1.2 Force Preparation—The ability to develop, enhance, adapt, and sustain the total force to effectively support national security.

1.2.1 Training—The ability to enhance the capacity to perform specific functions and tasks using institutional, operational, or self-development (to include distance learning) domains to improve the individual or collective performance of personnel, units, forces, and staffs.

1.2.2 Exercising—The ability to plan, prepare, execute, and evaluate maneuvers or simulated operations to validate training or conduct mission rehearsal.

1.3.1 Personnel and Family Support—The ability to provide the essential programs and services that support total force members and their families' quality of life and development in a transforming and expeditionary environment.

---

[1] A full list of Joint Capability Areas is available at http://dcmo.defense.gov/products-and-services/business-enterprise-architecture/9.0/reports/bealist_jointcapabilityarea_na.htm.

1.3.1.1 Community Support—The ability to sustain a military member and family support platform encompassing tuition assistance, children's education, spouse training and employment, child and youth services, morale welfare and recreation, and other programs that underwrite support to military members and their families.

1.4 Health Readiness—The ability to enhance DoD and the nation's security by providing health support for the full range of military operations and sustaining the health of all those entrusted to our care.

1.4.1 Force Health Protection—The ability to promote, improve, conserve, and restore the mental and physical well-being of deployed forces.

1.4.2 Health Service Delivery—The ability to provide acute or long-term primary or specialty-care capabilities to all eligible beneficiaries outside the theater in either the direct or purchased care system.

SOURCE: DoD, 2011b.

Relevant JCAs Associated with Battlespace Awareness (BA)

2. Battlespace Awareness—The ability to understand dispositions and intentions, as well as the characteristics and conditions of the operational environment that bear on national and military decisionmaking by leveraging all sources of information, to include intelligence, surveillance, reconnaissance, meteorological, and oceanographic.

2.1 Planning and Direction—The ability to synchronize and integrate the activities of collection, processing, exploitation, analysis, and dissemination resources to meet BA information requirements.

2.2 Collection—The ability to gather data and obtain required information to satisfy information needs.

2.3 Processing/Exploitation—The ability to transform collected information into forms suitable for further analysis and/or action by man or machine.

2.4 Analysis, Prediction, and Production—The ability to integrate, evaluate, interpret, and predict knowledge and information from available sources to develop intelligence and forecast the future state to enable situational awareness and provide actionable information.

2.5 BA Data Dissemination and Relay—The ability to present, distribute, or make available intelligence, information and environmental content, and products that enable understanding of the operational/physical environment to military and national decisionmakers.

SOURCE: DoD, 2011b.

Relevant JCAs Associated with Force Application

3. Force Application—The ability to integrate the use of maneuver and engagement in all environments to create the effects necessary to achieve mission objectives.

> **3.1 Maneuver**—The ability to move to a position of advantage in all environments to generate or enable the generation of effects in all domains and the information environment.
>
> > **3.1.4 Maneuver to Secure (MTS)**—The ability to control or deny (destroy, remove, contaminate, or block with obstacles) significant areas, with or without force, in the operational area whose possession or control provides either side an operational advantage.
>
> **3.2 Engagement**—The ability to use kinetic and nonkinetic means in all environments to generate the desired lethal and/or nonlethal effects from all domains and the information environment.
>
> > **3.2.1.3.1 Air**—The ability to kinetically engage moving targets in the region beginning at the upper boundary of the land or water and extending upward to the lower boundary of the Earth's ionosphere (approximately 50 km).

SOURCE: DoD, 2011b.

Relevant JCAs Associated with Logistics

4. Logistics—The ability to project and sustain a logistically ready joint force through the deliberate sharing of national and multinational resources to effectively support operations, extend operational reach, and provide the joint force commander the freedom of action necessary to meet mission objectives.

> **4.1 Deployment and Distribution**—The ability to plan, coordinate, synchronize, and execute force movement and sustainment tasks in support of military operations. Deployment and distribution include the ability to strategically and operationally move forces and sustainment to the point of need and operate the joint deployment and distribution enterprise (JDDE).
>
> > **4.1.1 Move the Force**—The ability to transport units, equipment, and initial sustainment from the point of origin to the point of need and provide JDDE resources to augment or support operational movement requirements of the Joint Force Commander (JFC).
>
> **4.2 Supply**—The ability to identify and select supply sources; schedule deliveries; receive, verify, and transfer product; and authorize supplier payments. It includes the ability to see and manage inventory levels, capital

assets, business rules, supplier networks, and agreements (to include import requirements), as well as assessment of supplier performance.

4.3 Maintain—The ability to manufacture and retain or restore materiel in a serviceable condition.

4.4 Logistics Services—The ability to provide services and functions essential to the technical management and support of the joint force.

> **4.4.1.4 Installation Feeding**—The ability to receive, store, prepare, and serve nutritious meals, authorized enhancements, and supplements in a sanitary dining facility at an enduring location or afloat based on Service ration cycle and mix, with the ability to project meals to disbursed populations.

> **4.4.3.3 Utility Operations**—The ability to manage and operate power, environmental control, water, and waste systems.

> **4.4.4 Hygiene Services**—The ability to provide laundry, shower, textile, and fabric repair support.

4.6 Engineering—The ability to execute and integrate combat, general, and geospatial engineering to meet national and JFC requirements to assure mobility; provide infrastructure to position, project, protect, and sustain the joint force; and enhance visualization of the operational area across the full spectrum of military operations.

> **4.6.3 Geospatial Engineering**—The ability to portray and refine data pertaining to the geographic location and characteristics of natural or constructed features and boundaries to provide engineer services. Examples include terrain analyses, terrain visualization, digitized terrain products, nonstandard tailored map products, facility support, and force beddown analysis.

4.7 Base and Installations Support—The ability to provide enduring bases and installations with the assets, programs, and services necessary to support U.S. military forces.

> **4.7.1.2 Facilities Support**—The ability to provide functional real property installation assets with utilities—energy, water, and wastewater; contract and real property management; pollution prevention; and essential services throughout natural or manmade disasters.

> **4.7.2.1 Security Services**—The ability to provide law enforcement (LE) functions and physical security to an installation.

>> **4.7.2.1.1 Law Enforcement**—The ability to provide the functions of LE operations.

>> **4.7.2.1.2 Base Physical Security**—The ability to provide physical security operations and support functions to safe-

guard personnel; prevent unauthorized access to equipment, installations/facilities, material, and documents; and safeguard them against espionage, sabotage, damage, and theft.

4.7.2.2 Emergency Services—The ability to protect and rescue people, facilities, aircrews, aircraft, and other assets from losses resulting from accident or disaster.

4.7.2.3 Installation Safety—The ability to prevent and respond to accidents and mitigate risk to the lowest acceptable level.

4.7.2.6 Airfield Management—The ability to provide airfield services, including weather, air traffic control, terminal/special use airspace management, airfield and flight management, cargo and passenger services, and transient aircraft support.

4.7.2.7 Port Services—The ability to perform and provide port services, including ship movements, berth days, magnetic silencing, cargo handling, transient vessel support, and waterborne spill response, at DoD and commercial seaports.

4.7.2.8 Range Management—The ability to safely maintain, schedule, control, and monitor ranges and uses associated with airspace/sea space and safety zone environments related to fixed point (nonmaneuver) ranges.

4.7.2.9 Launch Support Services—The ability to provide assistance for payload and launch vehicles including safety, reception, staging, integration, movement to the launch platform, and return to use activities after launch operations, at federal and commercial spaceports.

SOURCE: DoD, 2011b.

Relevant JCAs Associated with Command and Control

5. Command and Control—The ability to exercise authority and direction by a properly designated commander or decisionmaker over assigned and attached forces and resources in the accomplishment of the mission.

5.2 Understand—The ability to individually and collectively comprehend the implications of the character, nature, or subtleties of information about the environment and situation to aid decisionmaking.

5.2.2 Develop Knowledge and Situational Awareness—The ability to apply context, experience, and intuition to data and information to derive meaning and value.

5.3 Planning—The ability to establish a framework to employ resources to achieve a desired outcome or effect.

5.3.1 Analyze Problem—The ability to review and examine all available information to determine necessary actions.

5.3.2 Apply Situational Understanding—The ability to use synthesized information and awareness applicable to a given situation or environment to further understand the problem.

5.5 Direct—The ability to employ resources to achieve an objective.

5.5.1 Communicate Intent and Guidance—The ability to promulgate a concise expression of the operational purpose, assessment of acceptable operational risk, and guidance to achieve the desired end state.

5.5.1.5 Provide Warnings—The ability to communicate and then gain acknowledgment of dangers implicit in a wide spectrum of activities by potential opponents.

5.5.1.6 Issue Alerts—The ability to forewarn military decision-makers, operating location populations, and civilian authorities of immediate threats and other dangers.

SOURCE: DoD, 2011b.

Relevant JCAs Associated with Net-Centric (NC)

6. Net-Centric—The ability to provide a framework for full human and technical connectivity and interoperability that allows all DoD users and mission partners to share the information they need, when they need it, in a form they can understand and act on with confidence, and protects information from those who should not have it.

6.1 Information Transport—The ability to transport information and services via assured end-to-end connectivity across the NC environment.

6.2 Enterprise Services—The ability to provide to all authorized users awareness of and access to all DoD information and DoD-wide information services.

6.2.1 Information Sharing—The ability to provide physical and virtual access to hosted information and data centers across the enterprise based on established data standards.

6.2.2 Computing Services—The ability to process data and provide physical and virtual access to hosted information and data centers across the enterprise based on established data standards.

6.2.3.1 User Access (Portal)—The ability to access user-defined DoD enterprise services through a secure single entry point.

6.2.4 Position, Navigation, and Timing (PNT)—The ability to determine accurate and precise location, orientation, time, and course corrections anywhere in the battlespace and to provide timely and assured PNT services across the DoD enterprise.

6.3 Net Management—The ability to configure and reconfigure networks, services, and the underlying physical assets that provide end-user services, as well as connectivity to enterprise application services.

6.3.3 Spectrum Management—The ability to synchronize, coordinate, and manage all elements of the electromagnetic spectrum through engineering and administrative tools and procedures.

6.3.4 Cyber Management—The ability to assure network support for all DoD missions through the synchronization, deconfliction, coordination, and awareness of all elements of computer network operations.

6.4 Information Assurance—The ability to provide the measures that protect, defend, and restore information and information systems.

SOURCE: DoD, 2011b.

Relevant JCAs Associated with Protection

7. Protection—The ability to prevent/mitigate adverse effects of attacks on personnel (combatant/noncombatant) and physical assets of the United States, allies, and friends.

7.1 Prevent—The ability to neutralize an imminent attack or defeat attacks on personnel (combatant/noncombatant) and physical assets.

7.2 Mitigate—The ability to minimize the effects and manage the consequence of attacks (and designated emergencies) on personnel and physical assets.

7.2.1 Mitigate Lethal Effects—The ability to minimize the effects of attacks or designated emergencies that have the potential to kill personnel and destroy physical assets.

7.2.1.9 Natural Hazards—The ability to minimize the effects of natural hazards that have the potential to kill personnel and destroy physical assets.

7.2.2 Mitigate Nonlethal Effects—The ability to minimize the effects of attacks or designated emergencies that do not have the potential to kill personnel and destroy physical assets.

SOURCE: DoD, 2011b.

Relevant JCAs Associated with Building Partnerships

8. Building Partnerships—The ability to interact with partner, competitor, or adversary leaders, security institutions, or relevant populations by developing and presenting information and conducting activities to affect their perceptions, will, behavior, and capabilities to build effective, legitimate, interoperable, and self-sustaining strategic partners.

8.1 Communicate—The ability to understand, engage, develop, and present information to domestic partner audiences to improve understanding and to foreign partner audiences to create, strengthen, or preserve conditions favorable for the advancement of U.S. government interests, policies, and objectives through the use of coordinated programs, plans, themes, messages, and products synchronized with the actions of all instruments of national power.

SOURCE: DoD, 2011b.

Relevant JCAs Associated with Corporate Management and Support

9. Corporate Management and Support—The ability to provide strategic senior level, enterprise-wide leadership, direction, coordination, and oversight through a chief management officer function.

9.4 Acquisition—The ability to organize and execute the activities necessary to provide materiel for DoD operations.

9.4.4 Production and Lifecycle Acquisition—The ability to convert raw materials by fabrication into required weapons or systems, including production-scheduling, inspection, quality control, and related processes and tailored product support to achieve specific and evolving life-cycle product support availability, reliability, and performance parameters.

SOURCE: DoD, 2011b.

Bibliography

Bowman, Steve, "Homeland Security: The Department of Defense's Role," Washington, D.C.: Congressional Research Service, RL-31615, May 14, 2003.

Bryant, Benjamin P., and Robert J. Lempert, "Thinking Inside the Box: A Participatory, Computer-Assisted Approach to Scenario Discovery," *Technological Forecasting and Social Change,* Vol. 77, No. 1, 2010, pp. 34–49.

Bumiller, Elizabeth, "A Day Job Waiting for a Kill Shot a World Away," *The New York Times,* July 29, 2012. As September 17, 2012: http://www.nytimes.com/2012/07/30/us/drone-pilots-waiting-for-a-kill-shot-7000-miles-away.html?smid=pl-share

Davis, Paul K., *Analytic Architecture for Capabilities-Based Planning, Mission-System Analysis, and Transformation,* Santa Monica, Calif.: RAND Corporation, MR-1513-OSD, 2002. As of June 24, 2012: http://www.rand.org/pubs/monograph_reports/MR1513.html

Davis, Paul K., *Structuring Analysis to Support Future Decisions About Nuclear Forces and Postures,* Santa Monica, Calif.: RAND Corporation, WR-878-OSD, 2011. As of June 24, 2012: http://www.rand.org/pubs/working_papers/WR878.html

Davis, Paul K., *Lessons from RAND's Work on Planning Under Uncertainty for National Security,* Santa Monica, Calif.: RAND Corporation, TR-1249-OSD, 2012. As of August 14, 2012: http://www.rand.org/pubs/technical_reports/TR1249.html

Davis, Paul K., and Paul Dreyer, RAND's *Portfolio Analysis Tool (PAT): Theory, Methods, and Reference Manual,* Santa Monica, Calif: RAND Corporation, TR-756-OSD, 2009. As of December 20, 2011: http://www.rand.org/pubs/technical_reports/TR756.html

Davis, Paul K., Steven C. Bankes, and Michael Egner, *Enhancing Strategic Planning with Massive Scenario Generation: Theory and Experiments,* Santa Monica, Calif.: RAND Corporation, TR-392, 2007. As of August 9, 2012: http://www.rand.org/pubs/technical_reports/TR392.hmtl

Davis, Paul K., Russell D. Shaver, and Justin Beck, *Portfolio-Analysis Methods for Assessing Capability Options,* Santa Monica, Calif.: RAND Corporation, MG-662-OSD, 2008. As of December 20, 2011: http://www.rand.org/pubs/monographs/MG662.html

Davis, Paul K., Russell D. Shaver, Gaga Gvineria, and Justin Beck, *Finding Candidate Options for Investment: From Building Blocks to Composite Options and Preliminary Screening,* Santa Monica, Calif.: RAND Corporation, TR-501-OSD, 2007. As of June 24, 2012: http://www.rand.org/pubs/technical_reports/TR501.html

Davis, Paul K., Stuart E. Johnson, Duncan Long, and David C. Gompert, *Developing Resource-Informed Strategic Assessments and Recommendations,* Santa Monica, Calif.: RAND Corporation, MG-703-JS, 2008. As of August 9, 2012:
http://www.rand.org/pubs/monographs/MG703.html

Defense Science Board, "More Fight—Less Fuel," Report of the Defense Science Board Task Force on Energy Security, February 2008. As of December 1, 2011:
http://www.acq.osd.mil/dsb/reports/ADA477619.pdf

Department of Defense, JCA Background, undated. As of January 12, 2012:
www.dtic.mil/futurejointwarfare/strategic/jca_background.doc

———, *Quadrennial Defense Review Report,* September 2001. As of September 14, 2012:
http://www.dod.mil/pubs/qdr2001.pdf

———, *Quadrennial Defense Review Report,* February 2010a. As of December 1, 2011:
http://www.defense.gov/qdr/images/QDR_as_of_12Feb10_1000.pdf

———, *Joint Capability Management Plan (JCAMP),* 2010b. As of January 12, 2012:
https://acc.dau.mil/adl/en-US/347746/file/48928/Joint%20Capability%20Area%20Management%20Plan%20-JCAMP%20Final%20-%2027%20Jan%202010.pdf

———, *Joint Capability Area (JCA) 2010 Refinement,* Memorandum for the Vice Chairman of the Joint Chiefs of Staff, 2011a. As of January 12, 2012:
http://www.dtic.mil/futurejointwarfare/strategic/jca_approvalmemo.pdf

———, *Joint Capability Areas, JCA 2010 Refinement,* April 8, 2011b. As of December 1, 2011: http://www.dtic.mil/futurejointwarfare/strategic/jca_framework_defs.doc

———, *Base Structure Report, Fiscal Year 2012 Baseline,* 2013. As of January 28, 2013:
http://www.acq.osd.mil/ie/download/bsr/BSR%20Baseline%20FY2012%20Jan072013.pdf

Department of Energy (DOE), *Year in Review: 2011, Energy Infrastructure Events and Expansions,* April 2012a. As of July 27, 2012:
http://www.oe.netl.doe.gov/docs/2011-YIR-043012.pdf

———, *Hurricane Sandy Situation Report #20,* November 7, 2012b. As of November 9, 2012:
http://www.oe.netl.doe.gov/docs/2012_SitRep20_Sandy_11072012_1000AM.pdf

Eaton Corporation, *Blackout Tracker, United States Annual Report 2011,* Raleigh, N.C., 2012. As of June 24, 2012:
http://www.eaton.com/blackouttracker

Environmental Protection Agency, *Nellis Air Force Base, Nevada Success Story,* February 2009. As of September 12, 2012:
http://www.epa.gov/oswercpa/docs/success_nellis_nv.pdf

Erckenbrack, Adrian A., and Aaron Scholer, "The DoD Role in Homeland Security," *Joint Force Quarterly,* Issue 35, pp. 34–41, 2004. As of September 12, 2012:
http://www.dtic.mil/doctrine/jel/jfq_pubs/0935.pdf

Federal Energy Regulatory Commission (FERC) and North American Electric Reliability Corporation (NERC), "Report on Outages and Curtailments During the Southwest Cold Weather Event of February 1–5, 2011, Cause and Recommendations," August 2011. As of December 30, 2012:
http://www.nerc.com/files/SW_Cold_Weather_Event_Final_Report.pdf

Gompert, David C., Paul K. Davis, Stuart E. Johnson, and Duncan Long, *Analysis of Strategy and Strategies of Analysis,* Santa Monica, Calif.: RAND Corporation, MG-718-JS, 2008. As of June 24, 2012:
http://www.rand.org/pubs/monographs/MG718.html

Goss, Thomas, *Building a Contingency Menu: Using Capabilities-Based Planning for Homeland Defense and Homeland Security,* Master's Thesis, Monterrey, Calif.: Naval Postgraduate School, March 2005.

Government Accountability Office, *Defense Critical Infrastructure: Actions Needed to Improve the Identification and Management of Electrical Power Risks and Vulnerabilities to DoD Critical Assets,* Report 10-147, 2009a. As of June 24, 2012:
http://www.gao.gov/assets/300/297162.pdf

———,, *Defense Critical Infrastructure: Actions Needed to Improve the Consistency, Reliability, and Usefulness of DOD's Tier 1 Task Critical Asset List,* Report 09-740R, 2009b. As of June 24, 2012:
http://www.gao.gov/new.items/d09740r.pdf

———,, *Homeland Defense: DoD Can Enhance Efforts to Identify Capabilities to Support Civil Authorities During Disasters,* Report 10-386, 2010. As of September 12, 2012:
http://www.gao.gov/assets/310/302659.pdf

Groves, David G., and Robert J. Lempert, *A New Analytic Method for Finding Policy-Relevant Scenarios,* Santa Monica, Calif.: RAND Corporation, RP-1244, 2007. As of August 8, 2012:
http://www.rand.org/pubs/reprints/RP1244.html

Hines, Paul, Jay Apt, and Sarosh Talukdar, "Large Blackouts in North America: Historical Trends and Policy Implications," *Energy Policy,* Vol. 37, No. 12, 2009, pp. 5249–5259.

Institute of Electrical and Electronics Engineers (IEEE), *1366-2003 IEEE Guide for Electric Power Distribution Reliability Indices,* 2004. DOI: 10.1109/IEEESTD.2004.94548

Intergovernmental Panel on Climate Change (IPCC), *Managing the Risks of Extreme Events and Disasters to Advance Climate Change Adaptation,* A Special Report of Working Groups I and II of the Intergovernmental Panel on Climate Change, C. B. Field, V. Barros, T. F. Stocker, D. Qin, D. J. Dokken, K. L. Ebi, M. D. Mastrandrea, K. J. Mach, G.-K. Plattner, S. K. Allen, M. Tignor, and P. M. Midgley, eds., Cambridge, UK, and New York: Cambridge University Press, 2012.

Jackson, Brian A., *Marrying Prevention and Resiliency: Balancing Approaches to an Uncertain Terrorist Threat,* Santa Monica, Calif.: RAND Corporation, OP-236-RC, 2008. As of August 9, 2012:
http://www.rand.org/pubs/occasional_papers/OP236.html

Jackson, Brian A., Kay Sullivan Faith, and Henry H. Willis, *Evaluating the Reliability of Emergency Response Systems for Large-Scale Incident Operations,* Santa Monica, Calif.: RAND Corporation, MG-994-FEMA, 2010. As of August 8, 2012:
http://www.rand.org/pubs/monographs/MG994.html

Joint Defense Capabilities Study Team, *Joint Defense Capabilities Study, Final Report,* January 2004. As of June 1, 2012:
http://www.acq.osd.mil/jctd/articles/JointDefenseCapabilitiesStudyFinalReport_January2004.pdf

Knight, William, *Homeland Security: Roles and Missions for United States Northern Command,* Washington, D.C.: Congressional Research Service, RL-34342, January 28, 2008.

Lachman, Beth E., Kimberly Curry Hall, Aimee E. Curtright, and Kimberly M. Colloton, *Making the Connection: Beneficial Collaboration Between Army Installations and Energy Utility Companies,* Santa Monica, Calif.: RAND Corporation, MG-1126-A, 2011. As of June 24, 2012:
http://www.rand.org/pubs/monographs/MG1126.html

Lempert, Robert J., Benjamin P. Bryant, and Steven C. Bankes, *Comparing Algorithms for Scenario Discovery,* Santa Monica, Calif.: RAND Corporation, WR-557-NSF, 2008. As of August 8, 2012: http://www.rand.org/pubs/working_papers/WR557.html

Massachusetts Institute of Technology (MIT), *The Future of the Electric Grid*, 2011. As of December 14, 2011: http://web.mit.edu/mitei/research/studies/documents/electric-grid-2011/Electric_Grid_Full_Report.pdf

Min, Seung-Ki, Xuebin Zhang, Francis W. Zwiers, and Gabriele C. Hegerl, "Human Contribution to More-Intense Precipitation Extremes," *Nature,* Vol. 470, No. 7334, 2011, pp. 378–381.

Narayanan, A., and M. G. Morgan, "Sustaining Critical Social Services During Extended Regional Power Blackouts," *Risk Analysis*, Vol. 32, No. 7, 2012, pp. 1183–1193.

National Research Council (NRC), *Terrorism and the Electric Power Delivery System*, Washington, D.C.: National Academies Press, 2012.

Pepco, "Pepco Completes Full Restoration of Customers Impacted by Derecho," July 8, 2012. As of July 30, 2012: http://www.pepco.com/welcome/news/releases/archives/2012/article.aspx?cid=2088

Public Law 110-181, National Defense Authorization Act of 2008, U.S. Congress, 2008.

Public Law 112-81, National Defense Authorization Act of 2012, U.S. Congress, 2011. Smith, Rebecca, "Texas to Probe Rolling Blackouts," *The Wall Street Journal,* February 7, 2011. As of July 30, 2012: http://online.wsj.com/article/SB10001424052748703989504576128493806692106.html

Stockton, Paul, "Testimony of the Honorable Paul Stockton, Assistant Secretary of Defense Homeland Defense and Americas' Security Affairs, Department of Defense, Before the Subcommittee on Energy and Power, The Committee on Energy and Commerce, United States House of Representatives," May 31, 2011a. As of June 24, 2012: http://republicans.energycommerce.house.gov/Media/file/Hearings/Energy/053111/Stockton.pdf

Stockton, Paul, "Ten Years After 9/11: Challenges for the Decade to Come," *Homeland Security Affairs* 7, 10 Years After: The 9/11 Essays, September 2011b. As of December 19, 2012: http://www.hsaj.org/?article=7.2.11